Published and distributed by Richard Bennett Photography
PO Box 385 Kingston Tasmania 7051 Australia
Telephone: +61 3 6229 2559
www.richardbennett.com.au

First published in Australia 2006 by Richard Bennett

Copyright the Crown in Right of Tasmania and Richard Bennett

All rights reserved

Photography by Richard and Alice Bennett
Text by Bruce Montgomery
Colour photographs on Fujifilm
Film processing by Photoforce
Scanning by Photolith
Graphics by de Vos design
Printing by the Printing Authority of Tasmania

ISBN: 0-9578110-4-7

Every effort has been made to ensure that the information in this book is accurate at the time of going to press.
The publisher cannot accept the responsibility for any errors or omissions.

ISLANDS
OF TASMANIA

Dedicated to the four wonderful women in my life
– Susie, Lucy, Alice and Claire

RICHARD BENNETT

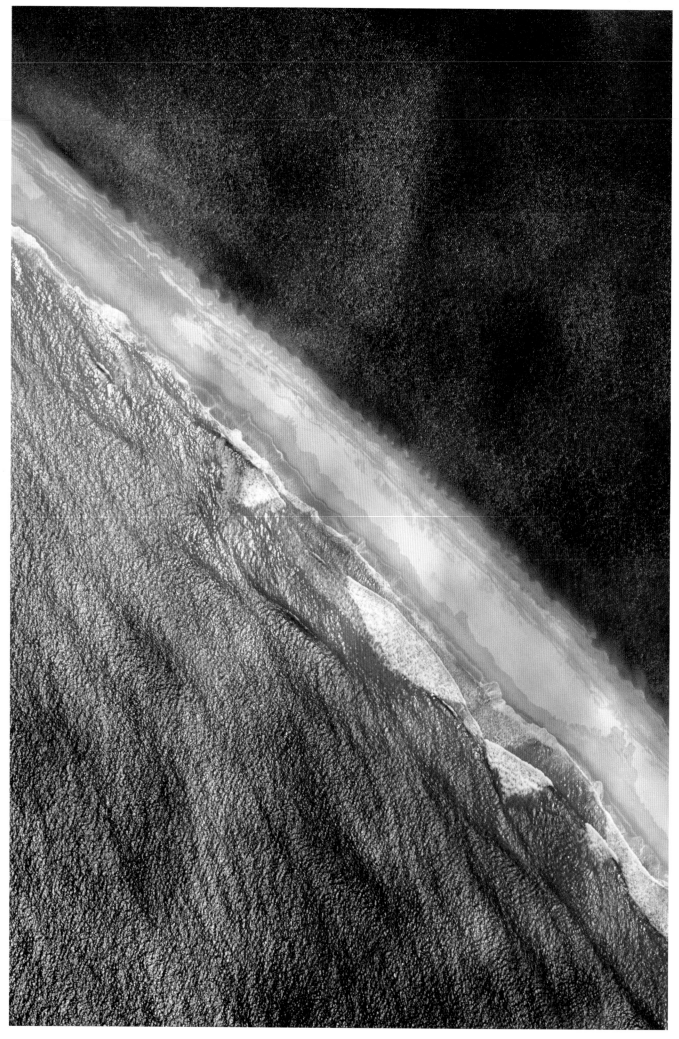

A beach separates the ocean and a lagoon at Cape Barren Island, east of Cone Point.

FOREWORD

The Honourable William J. E. Cox AC, RFD, ED – Governor of Tasmania

Bruce Montgomery, who has written the text for this outstanding collection of photographic studies of Tasmania and its many adjacent islands, rightly asserts that island people are different from others and have a sense of solidarity and a mutual bond comes with the territory of real or perceived isolation in their unique lifestyle. Shakespeare gave the concept expression through the mouth of John of Gaunt in his tribute to England:

> This happy breed of men, this little world,
> This precious stone set in the silver sea,
> Which serves it in the office of a wall
> Or as a moat defensive to a house,
> Against the envy of less happier lands;
>
> (Richard II, Act II, Scene I)

Tasmania and its islands are precious, as Richard Bennett demonstrates in this, his 11th book of photographs. From the crystal clear waters of the islands of the Furneaux Group and the East Coast of Tasmania with its orange lichen covered granite rocks, past the towering cliffs of the southern coast battered by the swell of the Southern Ocean, to the peat coloured waterways of the south-west he presents those images with which Tasmania and its people have come to be identified – purity of light, cleanliness of air, a pristine environment, a rugged independence of spirit and the capacity to develop high quality produce. The innovative use of Tasmania's natural resources combined with the outgoing, friendly and compassionate nature of its people has led to the expansion of markets for all manner of products of a temperate climate and a clean environment. It is not only a land of milk and honey but of much more – of fine wine and other beverages, fruit, timber, fish and all manner of agricultural products. Nor are its beauty and lifestyle conducive to tourism and bucolic pursuits alone, for it is an ideal location for scientific research, especially in view of its proximity to Antarctica, and for other intellectual activities.

Richard Bennett is passionate about our island state and brings his highly artistic skills to show why it is that Tasmania can claim to be envied by "less happier lands".

ISLANDS OF TASMANIA

Bruce Montgomery

Wherever you go in the world you will find that island people are different.

Whether they call the Orkneys or the Cook Islands their home, they have a sense of self-assurance about them, a sense of pride, a definite bond between themselves and a sense of solidarity. It comes with the territory of real or perceived isolation in an island lifestyle.

Tasmanians fit that island mould. They may argue among themselves in a parochial sense, a town versus town or region versus region thing, but when it comes to the crunch they are Tasmanians first, Australians second and northerners or southerners last. It makes for a warm, outgoing people, characterised by an over-endowment of generosity and an elevated sense of self-sufficiency.

With some island people that can verge on a sense of insularity, often the downside of living on an island, the tendency to have a narrow and ungenerous view of the world because of a mindset that your own backyard is all the world you need.

While it may have applied in the past in Tasmania, it clearly no longer is the case. Part of the explanation is that it is so much easier and cheaper to travel, to get off the island. Tasmanians know they are privileged to live in one of the most beautiful places on Earth: an uncrowded island, largely unspoiled, a temperate climate, prosperous, ingenious and generous people. Our shores are washed by clean seas. Our air is the cleanest on Earth. We are a sanctuary at the bottom of the world.

When Tasmanians return from the far corners of the world, they step onto a land where you can still see for miles and miles in a clear light, where the people are genuine, perhaps sometimes unsophisticated, but genuine, and where there is an overwhelming sense of security and that here life is lived as it should be.

Tasmanians live in a stable democracy. They have perhaps the most democratic form of electing their governments, the Hare-Clark system of proportional representation that produces MPs in direct proportion to each party's share of the vote.

There is a definite sense of "Tasmanian-ness" here. It is manifested in how Tasmanians relate to each other, in the ease with which they live with their environment and the sense of self-sufficiency that they have from having lived their lives on an island. On an island you can't always depend on someone else. You have to do things yourself.

On Tasmania's offshore islands, places like Flinders, King and Bruny, those characteristics are even more pronounced.

It is on these islands where you see the real grandeur of this place – clear water; pristine, deserted beaches; abundant sealife; and small communities comfortable with themselves.

Cone Point on Cape Barren Island with Passage, Forsyth and Clarke islands in the distance.

Tasmania was not always one island, surrounded by another 330 or so. It was not an island at all.

About 550 million years ago it was part of Antarctica, part of the supercontinent of Gondwana, which was composed of present-day South America, Africa, India, Australia and Antarctica.

Some of our endemic plants today reflect our Antarctic lineage. The scientific name of the Tasmanian tree fern is *Dicksonia antarctica*. Our myrtle beech (*Nothofagus cunninghamii*) and sassafras (*Atherosperma moschatum*) are Gondwanan. Huon pine (*Lagarostrobus franklinii*) is a living fossil from those times.

Gondwana started to disintegrate about 130 million years ago. The Australian continent formed as a separate entity, moving north, away from Antarctica and separated from it by the circumpolar current. Australia evolved into a warm and wet place, dominated by rainforest, then to a hot and arid land.

About 18,000 years ago, yesterday in geological terms, the level of the sea around Tasmania was 120 metres lower than today. At this time there was a land bridge between Tasmania and the Australian mainland. It allowed the free travel of Tasmanian Aborigines via what is now Flinders Island.

When sea levels rose at the end of the last Ice Age, finishing 11,000 years ago, the present configuration, a main island with an archipelago of islands, was created by the inundation caused by the thaw.

Today the islands extend from 10 kilometres off the Victorian coast to Macquarie Island, 1500 kilometres to the south-east in the sub-Antarctic.

Night Island

Night Island

Night Island is an important colony for seabirds, including Australian pelicans.

Giant granite boulders on Night Island, westernmost of the islands in the Preservation Island group.

Preservation and Rum islands. Rum Island is where the survivors of the *Sydney Cove* stored their supplies of rum after she foundered. Armstrong Channel, beyond, has strong tides and a shallow, sandy bottom, which makes for an interesting navigational exercise.

Dawn on Preservation Island, upon which the *Sydney Cove* ran aground in February 1797 en route from Calcutta to Port Jackson.

The catamaran *Resolution* anchored off Preservation Island.

Cape Barren goslings on Preservation Island. The Cape Barren goose (*Cereopsis novaehollandiae*) has the rare ability to drink salt water. It is a grazing bird, now more abundant than at any other time since European settlement. The geese are found as far west as the Recherche archipelago near Esperance, WA.

Preservation Island

Clarke and Forsyth islands

On Easter Island such figures were carved by early Man. On Clarke Island nature shaped them.

Clarke Island

Clarke Island's Snug Cove

Winds of more than 50 knots sweep across Cape Barren's tussocks towards Vansittart Island at the eastern end of Franklin Sound.

PREVIOUS PAGES: Cape Sir John on Cape Barren Island. Originally there was a land bridge between Tasmania and the Australian mainland. It was engulfed by the waters that are now Bass Strait about 11,000 years ago. The shallow, narrow waterway can be characterised by short, steep waves during heavy winds, especially the westerlies that flow along these latitudes. Cape Sir John, the westernmost point of Cape Barren Island, is often the first landfall to feel them.

A bush chair made by the Tasmanian artist and sculptor Gay Hawkes, whose specialty is objets d'art using found and recycled materials. She produced this chair with students on Cape Barren.

Casuarina, Cape Barren Island

One of Tasmania's virtues is its clear, sharp light. The lack of air pollution means you can see for ever.

Casuarinas and granite

Franklin Sound

The wreck of the windjammer *Farsund* off the south-east coast of Vansittart Island at the eastern entrance to Franklin Sound. She was driven aground in a gale in March 1912 on a voyage from Buenos Aires. Vansittart was a favourite haunt of sealers in the early 19th century.

Shai Maynard, one of the younger generation of Tasmania's Aboriginal people who continues the traditional hunting practices of their ancestors.

Aborigines first inhabited Tasmania about 30,000 years ago. It made Tasmanian Aborigines the most southerly peoples in the world during the Pleistocene era, from about one million years ago to 11,000 years ago.

With the flooding of the land bridge and the neighbouring plains by the waters that are now Bass Strait, the Aborigines evolved into a seafaring people. They crafted canoe rafts to begin hunting the seal colonies on the west and south-east coasts. They harvested crayfish, oysters, scallops, mussels, limpets, abalone and seabirds' eggs. They caught penguins, muttonbirds and seals.

The new islands were no barrier to them. The early European explorers observed Aborigines on Bruny, Maria and Schouten islands.

The islands of the Furneaux Group at the eastern end of Bass Strait were named by the English explorer Tobias Furneaux, who sighted them in 1773. They hold a special place in the hearts of Tasmania's Aborigines. In the 1830s and 1840s the remnants of the Aboriginal population after white settlement were transferred from the Tasmanian mainland to Flinders Island. Today 1000 people, many of them descendants of the original people, live in the Furneaux Group.

In the early days of white settlement, these islands were the domain of white sealers, who called themselves "Straitsmen". They were a law unto themselves. They were on the islands for one purpose – to plunder the abundant populations of seals.

Sealing and whaling had been major industries in Tasmania even before white settlement in 1803. Captain Charles Bishop in the *Nautilus* established a base on Cape Barren Island and collected more than 10,000 seal skins in five months.

Seal numbers soon declined and the big operators turned to whaling. By the 1830s, with the arrival of free settlers and merchants, whaling accounted for 25 per cent of the colony's exports. By 1850, 40 whaling vessels from 100 to 200 tons in size were registered in Hobart, with 1000 employed in the industry.

In March 1826 a writer to the *Hobart Town Gazette* said:

"The unthinking sealers harass these useful animals at all seasons, and the consequence is that many reefs are deserted, and inferior skins have been procured from animals too young, and imposed upon the merchants."

In 1871 Charles Gould wrote in the Royal Society of Tasmania's journal:

"In continuation of my remarks upon the islands in Bass' Straits (sic), I have now to advert to the Mammalia and, more especially to pen the deplorable memento of the gradual decrease, and now rapidly approaching extinction of some of those species adapted by form and structure to inhabit the wildest and least accessible spots, and whose abundant presence formerly on the detached reefs and rocky coasts throughout the group must have imparted a gratifying air of animation to what are now dumb and barren solitudes.

"It is well, I think, for naturalists to begin to assemble the pages of the history of a species without waiting until the tombstone has been finally erected above it, and I have therefore gathered from the journals of the earlier voyages such incidental remarks as illustrate the aspect of the island in point of the abundance of the amphibious carnivore, prior to the hostile invasion of man.

"It seems hardly credible that wanton apathy should have permitted a wholesale extermination at all seasons of so valuable an article of commerce; but true it is that no steps appear ever to have been taken to afford protection to the various species of seal during the fence season, and the inevitable result of so persistently ungenerous a persecution, has been their almost total disappearance from localities once abounding in them by thousands."

Gould said there was evidence that, in one instance, 300 seal cubs on one bank alone had perished because their mothers had been clubbed to death for their five-shilling skins.

Great Dog Island, which is more commonly known locally as Big Dog, with Vansittart in the distance.

Muttonbirding on Big Dog Island

Muttonbirders Rex Johnson, Robert Hall and Craig Broadbent on Big Dog. The short-tailed shearwater is known as the muttonbird. The birds nest in burrows on Tasmania's islands and along the coastline of the Tasmanian mainland. The chicks are a delicacy of Tasmanian Aborigines; they call it *yolla*. The season lasts for a little over one month and is very much a family business. The oil from the young birds is used for pharmaceutical goods.

Bernice Condie of Lady Barron makes necklaces from small maireener shells that she harvests from the waters around the islands in Franklin Sound. She dries, cleans and threads the shells and sells them through a Hobart gallery. The maireener shell necklaces are a traditional form of decoration of the Tasmanian Aboriginal people.

The town of Whitemark and the Strzelecki Range on Flinders Island. The largest island in the Furneaux Group, Flinders was a significant part of the land bridge that existed between Tasmania and the mainland of Australia. The island was settled by sealers and is now used for agriculture, with much of the land having been divided up into soldier settlement blocks in the 1950s. It has some of Tasmania's best dairy herds.

In 1833 those surviving Tasmanian Aboriginal members remaining on the Tasmanian mainland, about 160 people, were transported to Settlement Point on the west coast of Flinders Island, a place they called Wybalenna. The relocation was said to be for their protection, but it signalled their demise. The chapel at Wybalenna and the adjacent cemetery are places of great significance to the Aboriginal community. The national inquiry into the Stolen Generation took its first evidence in the chapel in December 1995.

The chapel at Wybalenna

Local naturalist and artist Fiona Stewart with her young wombat

The Flinders Island Soldier Settlement War Service Scheme, established after World War II, changed life on Flinders.

It began in 1951, allocating undeveloped but potentially productive land to servicemen from the island. It was then made available to servicemen across the country, bringing an influx of new settlers. The population doubled from 600 people in less than a decade.

Fisherman Tony Harper with a cage full of good sized crays at Lady Barron

This is as remote as it gets in Tasmania. The lighthouse on the southern tip of Goose Island warns shipping approaching from the west of the proximity of islands and shoals in the sound. It was built by convicts in 1846. It had a full staff of head keepers and assistant keepers and their families who lived in cottages that have since been demolished. There is a cemetery on the island, but the headstones of keepers are showing signs of the years of exposure. After being automated, the lighthouse was converted to wind power in 1985 but there were continual maintenance problems because the winds proved so fierce. It was eventually converted to solar power.

East Kangaroo Island, south-west of Whitemark, has been used extensively for grazing.

James Luddington skippers the charter boat *Strait Lady*, which works out of Lady Barron.

The east coast of Flinders has a series of lagoons and inlets that are important sanctuaries for waterfowl, including the Australian pelican.

Castle Rock on the west coast of Flinders

Swedish-born former merchant seaman Arne Eriksson has built a village from driftwood and the flotsam and jetsam of life that washes up on the coast of Flinders. His creations are now one of the island's attractions. Some of generous disposition have compared Eriksson's work with that of the Spanish architect Antoni Gaudi. You be the judge.

Arne with neighbour Daryl Butler

The west coast of Flinders has many secluded beaches with four-wheel drive access, making them ideal for private picnics.

Erith Island in the Kent Group in mid Bass Strait with Deal Island across the passage known as Murray Pass. The small island is North East Island. For years, Melbourne writers Stephen, Nita and Joanna Murray-Smith spent time on the island.

"From our island, we could look across Murray Pass, the dangerous stretch of water between us and Deal, towards human habitation," Joanna wrote. "At night, the triple beam of the lighthouse was our comfort against the darkness and loneliness of the strait.

"From Erith, the syncopated light – three turns, the pause, then the three rotations again – was a comforting antidote to our remoteness."

Her book, *Judgement Rock*, was set on Deal.

To the people of Tasmania's Bass Strait islands, Hughie Sinclair was sometimes the hero, sometimes the villain, always the lifeline.

Hughie died in 2004. He had been the pilot, the mail man, the grocer and the purveyor of news both factual and alleged to the thousand or so people who live in Tasmania's most remote communities at this eastern end of Bass Strait. Hughie was a daredevil, a renegade and a bit of a rogue.

The islands are only an hour's flying time from Melbourne. The lifestyle they offer is light years from modern Australia.

I spent some time with Hughie, flying the mail run around the islands from his base at Lady Barron.

He picked me up early one morning at the airstrip at Bridport on the north Tasmanian coast. Another passenger was waiting. Barney Howell, a drover, was sitting in the shade of a tree at the edge of the airstrip. Beside him were his two dogs. They were alert, clever dogs. They sniffed the breeze at the beginning of this hot summer day. They sensed that today would be different to most other days of their lives. When Hughie Sinclair's Cessna came into view and landed in the nearby strip, the dogs were on their feet. Today was to be different.

Hughie alighted from his plane. He had a full mane of greying hair, a relaxed manner about him, no airs and graces, a man who had found his natural home around Franklin Sound, the shoal-ridden strait between Flinders and Cape Barren islands. Hughie had a sense of mischief about him. It was in his eyes. He challenged you to take him on, work out whether he was serious.

Hughie had been expelled from school for "aberrant behaviour". He dragged himself up by the bootstraps and made something of himself. Before he became a pilot, he was in the State Government service. He was like a square peg in a round hole. He didn't like rules.

When you flew with Hughie, the only visual indication that he was a pilot was the epaulettes on his shoulder, a quasi-military adornment that did not quite fit the image.

That morning we were to drop Barney and the dogs on Waterhouse Island, north-east of Bridport. Barney had been contracted to drench the 1300 sheep there, grazing unattended with the resident herd of deer and the island's flock of Cape Barren geese.

It was a 10-minute flight. There was only need to climb to 2000 feet. The dogs were in the back seats, looking out the window, unflustered by the din that the Cessna was making. It was too loud to converse properly. Ned, the elder of the dogs, decided to come up and see Hughie. He jumped up on his lap for a spell, got bored and went back to his seat.

On these islands, like most airstrips in Tasmania's outlying places, there is no fencing to keep the native and domestic animals away from the strip. You have to buzz the landing area first to scare the pants off the local animals, drive them away long enough to have a clear run at landing. The deer, the geese and the sheep all duly scattered and Hughie landed. Barney and the two dogs climbed out.

"See ya, Barney."

"See ya, Hughie."

They took off and so did Hughie.

Up until shortly before he died, Hughie flew his mail run to all of the islands where he could land – Cape Barren, Clarke, Swan, Deal, Passage, Preservation, Badger and Chappell.

Chappell Island is the most feared in the strait. It has a dense, in-bred population of tiger snakes. They feast on muttonbird chicks for six weeks a year and fast for the rest. Passengers in transit choose to stay on the plane when it lands on Chappell Island.

On Cape Barren Island, a group of people, most of them Tasmanian Aborigines, was waiting at the landing strip and soon surrounded the door to the freight compartment of Hughie's plane. The shop took off its bread, apples, milk, magazines, a tin of black stove polish, whatever else has been ordered.

On any day Hughie would carry boxes of food and grog, spark plugs, new tyres, magazines educational and erotic, and a steady parade of islanders, to and from medical appointments on Flinders or in Launceston and even on the one-way flight to the undertakers.

At the end of my jaunt around the islands with Hughie, I told him I couldn't recall him having radioed (to air traffic control) in at any stage.

"You couldn't have been listening," he said.

Hughie was one of the few pilots skilled enough to land on Deal Island in the Kent Group. The strip is high on the face of the island and has a dog-leg two-thirds along. Landing is interesting; taking-off is like launching a hang glider.

Deal Island is half-way between Flinders Island and Wilsons Promontory. It is a favourite stopover for yachts crossing Bass Strait. There are anchorages that are sheltered from most weather.

The Deal Island lighthouse was turned on in 1848. It was built jointly by the New South Wales, Victorian and Tasmanian governments, using convict labour. At an elevation of 305 metres, it was the highest light in the Southern Hemisphere.

The tower was constructed of rubble using local granite. Most other materials had to be brought to the island and hauled up to the site three kilometres from the landing. The construction was very labour intensive. The construction crew was made up of an officer in charge of the convicts, an overseer, a medical dispenser, two carpenters, four masons, two quarrymen, a blacksmith and 19 labourers. They used 10 bullocks to haul the construction materials.

The elevation of the light often caused visibility problems. Low cloud obscured the light 40 per cent of the time, but the keepers at Wilsons Promontory lighthouse 75 kilometres to the north reported being able to see the Deal Island light on an average of six nights in 10.

Today the light offers no warning to passing mariners. It was switched off in 1992.

The first lighthouse keeper is a well-known name around the Tasmanian coast, William Chaulk Baudinet, who arrived in Hobart with his French wife Augusta Louisa Jane Baudinet in 1830. They first served at the Bruny Island light. They had 12 children (which may be testament to the attraction of lighthouse keeping). At one stage nine were said to be with them on Deal. William stayed on until his death in 1865. Four Baudinet children became lighthouse keepers. Charles Baudinet became keeper of the Swan Island light in 1867, staying there until his retirement in 1891.

The future of the island is uncertain. In 1998 the Australian Maritime Safety Authority handed it over to the State Government. The Australian Bush Heritage Fund, which also had a lease on nearby Erith Island, had a short-term lease. The Tasmanian Parks and Wildlife Service supervises a program of resident volunteers on the island.

Deal Island, with the lighthouse on the headland and the keepers' cottages on the plain

Rodondo Island is just 16 kilometres off the coast of Victoria's Wilsons Promontory but is part of Tasmania and is its northern extremity. From cliffs 200 metres high, Rodondo soars to 350 metres at its summit.

Rodondo Island

Devils Tower, halfway between the Kent Group in Bass Strait and Wilsons Promontory in Victoria, is home to colonies of short-tailed shearwaters, fairy prions, diving petrels and Australian fur seals. That should be sufficient cause for it not to be used for target practice by the military. According to Nigel Brothers' book, *Tasmania's Offshore Islands*, the island, in fact two islets, is an authorised military target.

Low Head lighthouse at the mouth of the Tamar River where it enters Bass Strait. There has been a signal station at Low Head since 1805, the first days of European settlement in the north. The original lighthouse was designed by the colonial architect John Lee Archer and built in 1833, with the crown made of Launceston freestone. It was replaced in 1888 with the present structure, which is of double brick. The lighthouse had Tasmania's only foghorn from 1929 until 1973. The light has been automated.

Waterhouse Island

The Dutch, the French and the British explored and named Tasmania's islands. Abel Tasman sighted the west coast on 24 November 1642 and named the main island Van Diemens Land in honour of Antony van Diemen, Governor-General of the Dutch East India Company. He did not stop there.

After making landfall on the west coast, he sailed south, observing and naming Maatsuyker, De Witt, Pedra Branca, Boreel and Tasman islands before sailing up the east coast and naming Maria and Schouten islands.

Tobias Furneaux, when his ship *Adventure* separated from James Cook's *Resolution* during his British expedition of the Southern Ocean, explored the east coast and named the Furneaux Group, but never deduced there was a strait between Van Diemens Land and the Australian mainland, then known as New Holland.

In 1798 George Bass and Matthew Flinders began a circumnavigation of Van Diemens Land in the sloop *Norfolk*, proving that it was an island. Flinders arrived back in Sydney in January 1799 having charted about 500 miles of the coast and a number of Bass Strait islands. His charts were not published until 1814.

In 1802 Frenchman Nicholas Baudin, in *Le Geographe* and *Le Naturaliste*, anchored off Bruny Island, before exploring D'Entrecasteaux Channel and the south-east and east coasts of Tasmania between Cape Pillar and Freycinet Peninsula. Freycinet was the second in command of *Le Geographe*. King Island is at the western entrance to Bass Strait. Today it has an international reputation for its gourmet foods and cuisine. For most of its history it was best known for its wrecks. King Island is a ships' graveyard.

The west coast of the island is in the path of the westerly winds that race along the latitudes known as the Roaring Forties. The combination of unrelenting winds and the lee shore that King Island presented to them meant it probably has more wrecks per kilometre of shoreline that anywhere else in Australia.

The most famous is that of the *Cataraqui*, which foundered in August 1845 with the loss of 400 lives. She left Liverpool for Melbourne with emigrants on board. Close to the end of her journey, she heaved to in a gale, but ran aground and quickly broke up.

One of the theories about why the pastures of King Island produce such fine beef and dairy foods is that the local grasses are derived from the straw mattresses made from English grasses washed ashore from shipwrecks.

Robbins Island, just off the coast of the far north-west, is privately owned by the Hammond family – Keith, John and Chauncey. By Tasmanian standards it is big. Robbins and its northern neighbour, Walker Island, cover 100 square kilometres. The Hammonds breed Japanese Wagu cattle on the islands, animals that produce the highly-marbled Kobi beef.

Keith Hammond says the Japanese like the Robbins Island Wagu beef because it comes from a clean environment, with clean water, clean air, and the cattle are hormone-free.

The Hammonds say they hold Robbins Island in trust.

"Our great-great grandfather came here in the 1850s from England and it's been in the family since 1916," John Hammond says.

"We feel a real bond with the land. We feel our duty is to pass it on to the next generation in the best condition we can, and even though we may say we own the land we are really only custodians during our lifetime."

Surely the most compelling and mysterious of the larger Tasmanian islands is Maria, off the east coast. It is an island of mystery, with a deeply-layered, textured history. Its cliffs are impregnated with fossils from the last Ice Age.

There is evidence of 30,000 years of Aboriginal occupation.

Maria was home to the Tyreddeme people, a band of the Oyster Bay tribe, who were among the southernmost people in the world after the last Ice Age.

The Tyreddeme occupied Maria at the time of colonisation, travelling there from the Tasmanian mainland on canoe rafts.

They crossed to the island regularly and used the ochre from Bloodstone Point near the isthmus to decorate their bodies and hair and to produce bark paintings. They built huts, hunted and gathered food. They disposed of their dead there in a very unusual way. In the Baudin expedition of 1802, zoologist Francois-Auguste Peron described vertical bark tombs, set out on poles and bound at the top to give them an appearance of a four-sided pyramid or a teepee. Inside were the ashes of humans.

The island was sighted and named by Tasman only days after he had made landfall on the west coast and named the main island Van Diemens Land. He named it Maria Eylandt after Maria van Diemen, the wife of Antony van Diemen.

In the ensuing years Maria Eylandt would seduce the French and the British. It was home to the Irish nationalist prisoner William Smith O'Brien (Maria Island had convicts before Port Arthur) and an enterprising Italian industrialist named Diego Bernacchi.

There were two periods of penal settlement on Maria, from 1825-30 and from 1842-50. Eventually there were up to 627 convicts on the island. In November 1849, Smith O'Brien, convicted of high treason, was brought to the island.

Diego Bernacchi was an Italian silk merchant who wanted to raise silk worms in Australia. He expected to bring 40 to 50 Italian families to the colony. He arrived in April 1884, aged 30, with a shilling-a-year lease for the island, but committed to spending £10,000 over 10 years.

He established 40 acres of vineyards, orange, lemon and mulberry plantations. There were fields of figs, mandarin trees, pomegranates, lemons, chestnuts and poplars.

In 1887 the Maria Island Company Limited was floated for fruit and wine growing, sheep and cattle raising, limestone and marble quarrying, agriculture, cement, timber, a sanatorium, land allotments and a fishery. The population rose to 250.

In 1888 Bernacchi renamed Darlington San Diego. The town had the Grand Hotel, shops, butcher, baker, blacksmith, shoemaker and post office. In the same year a coffee palace was built on the site of the separate apartment cells, using the bricks.

In 1889 work started on the cement works east of Darlington, but three years later the Maria Island Company was placed in liquidation.

Bernacchi returned in 1920, bought the township and a 400,000 tonne limestone deposit. He completed the cement works and a wharf to export the cement. But the limestone was contaminated by impurities and the venture failed.

In 1972 Maria was proclaimed a national park. Its separation from the Tasmanian mainland serves a 21st century purpose. It has made Maria a modern-day Noah's Ark for Tasmania's threatened species. In the past 25 years a number of threatened species have been introduced here in a bid to build their numbers.

The characteristics that made the island a convict settlement now make it an ideal refuge for plant and animal species that are elsewhere under threat. The rare forty-spotted pardelote is a local bird found here in good numbers. In 2005 nine Tasmanian devils were moved to Maria in a bid to protect them from a facial tumour disease.

From Cape Tourville looking south-west across Sleepy Bay towards Mount Graham and Mount Freycinet and Wineglass Bay in the Freycinet National Park

A pod of pilot whales in distress on the beach at Darlington on Maria Island in November 2004

The isthmus between the two halves of Maria Island with Riedle Bay on the left, Shoal Bay in the centre and Booming Bay on the right

Marion Bay

At the end of the year, thousands of music lovers flock to an East Coast farm that overlooks Marion Bay and Maria Island for the Tasmanian Falls Festival. This is Woodstock without the mud.

The dramatic cliffs of Tasman Island on the south-eastern corner of Tasmania. Tasman is a mark of the course of the annual Sydney-Hobart yacht race. The lighthouse was built in 1906. It is constructed of cast-iron plates bolted together on top of a concrete base 26 metres in diameter. Tasman Island was once thickly forested. The brick cottages on the island were also built in 1906. Stores were delivered to the lighthouse keepers using a haulage way and flying fox. In one famous storm in 1919 verandahs and fences were destroyed, water tanks blown off their stands and outbuildings moved on their foundations. The light was automated in 1976 and de-manned in May 1977. The wind generator was replaced by solar power in 1991.

The Penitentiary. Originally established as a timber station, Port Arthur became a prison settlement for male convicts in 1830 and operated until 1853. It housed about half the convicts transported from England to Australia, 1200 at any one time. An estimated 12,500 convicts passed through Port Arthur.

Cape Raoul. Tasmania's rugged terrain provides a dramatic and challenging environment for pilots.

The Iron Pot lighthouse marks the mouth of the Derwent River. It was the first lighthouse to be built in Tasmania and is Australia's second oldest. The structure that you see today is the remnant of an impressive collection of buildings that was on the small island. There were several keepers' cottages and outbuildings. The existing tower began operating in 1833. A big storm devastated the buildings in 1895. Kelp was said to be clinging to the uppermost rails of the lighthouse. The keepers were withdrawn in 1921, the cottages dismantled and removed to Hobart.

Deep water, even close to the shore, guarantees some of the best sailing in Australia.

In December 2004 the French yachtsman Roland Jourdain limped into Hobart after abandoning his attempt to win the Vendee Globe solo, non-stop, round-the-world race when the keel of his boat was damaged.

Hobart is a haven for seafarers completing circumnavigations of the world. It is the halfway point in the transit of the Southern Ocean and the psychological halfway point of the entire voyage.

Each year local yachtsmen and women compete against the world's best in the Sydney-Hobart yacht race. Size and age is of no importance. To have competed and completed is the honour.

Tasmania's capital city, Hobart

Summer Festival street performance on the
Hobart waterfront

Hobart is a maritime city, renowned throughout the world for its
stunning scenery and hospitality. Here there is a tradition of the
sea and of things nautical. Each year, just after Christmas, Hobart
comes alive with the arrival of yachts competing in the Sydney-
Hobart and the Melbourne-Hobart races. To coincide with their
arrival the Taste of Tasmania food festival showcases the best of
Tasmanian food and wine and also brings out the best of the visual
and performing arts.

Tasmanian Symphony Orchestra

Salamanca night life

Battery Point terraces

A great start on the beach at Blackmans Bay

Kettering

This juvenile sea eagle is one of the lucky ones. It was injured in a collision on the East Coast. It damaged its tail feathers and was unable to gain altitude. The bird found its way into the hands of Craig Webb, who runs the Raptor and Wildlife Refuge at Kettering. The eagle stayed with Webb at the refuge's aviary for 19 months while it grew new tail feathers. Then, on one fine day in D'Entrecasteaux Channel, Webb released it back into the wild.

"It landed on the branch of a dead tree and stayed there for the next four hours before taking off," he said.

The Neck on Bruny Island, the isthmus that connects North and South Bruny

The coast of North Bruny with Adamsons Peak in the background

Adventure Bay

Shack at Miles Beach

Great Bay on Bruny Island

Grand designs. Pilot Ralph Schwertner and Teresa Derrick with their shack nearing completion at Sheepwash Bay.

Lunawanna, Bruny Island

Cape Bruny

Andy and Beth Gregory epitomise island people. They are lighthouse keepers who no longer have to keep the light, but they remain with the one where they have spent much of their lives, Cape Bruny on Bruny Island. It is said that Andy never goes anywhere without a tin of Brasso in his pocket.

The Gregorys could lay claim to being Australia's most remote residents. Bruny is an island off an island off an island.

They have been there full time since 1995, but first manned the light in 1983 at a time when lighthouse keepers worked the circuit of Tasmanian lights.

Today Tasmania's lights have been automated. For most, that means no human presence, no eye scanning the ocean in the middle of the night. Andy still reports visual sightings to the Bureau of Meteorology. When you hear there is a two-metre south-westerly swell at Cape Bruny, it will have been Andy who saw it.

The Cape Bruny lighthouse is a classic stone tower, painted white, designed by the colonial architect John Lee Archer.

It was built by 12 convicts in 18 months, completed in 1838. Flanked by the great southern capes at the bottom of the world, it stares out at the Southern Ocean to Antarctica beyond the horizon.

For the convicts and settlers who sailed out from England in the 19th century the white flash once every 50 seconds from the Cape Bruny light was likely their first sight of land and civilisation since leaving England many months before.

The Baudinets, whom we have already met in the section on Deal Island, raised their brood of children at Cape Bruny.

Today the lighthouse is in perfect working order but since 1996 a modern beacon powered by the sun has taken its place.

An isthmus halfway down the island divides it into two quite different landscapes, each reflecting the increasing dramatisation of the Tasmanian mainland landscape the further south one gets.

The island and its dividing channel take their names from the French explorer Bruni D'Entrecasteaux.

Bruny is home to another remote Australian, former Sydney brain surgeon Dr Ian Johnston. He is a noted neurosurgeon, an expert in the causes of fluid pressure on the brain and the surgical treatment of epilepsy.

But, at the age of 60, he turned his back on neuroscience and the study of cerebrospinal fluid circulation to concentrate on his passions, the translation of Chinese literature and classical Greek texts. He has degrees in both, and a PhD in Mandarin. His home is on 20 hectares of land nestled in dunes behind Cloudy Bay at the southern end of Bruny.

Each room in his small house at Cloudy Bay, apart from the kitchen, is lined with bookshelves filled with volumes of Chinese and Greek texts. He gravitates from one room to another as the sun moves across the sky. He works through the night on his translations, a sole light burning on storm-swept nights when the southerlies whip up from the Antarctic to thrash the bay.

This place, on an island off an island off an island, is his seclusion.

Cloudy Bay Lagoon

Hobart artist Roger Murphy completes one of his watercolours at Partridge Island.

Beach at Cloudy Bay

Huon River at Franklin

Rowing on the Huon at Egg Island. The Huon has produced many great international rowers, including Athens Olympic silver medallist Simon Burgess.

Arve Falls

Kelvin Aldred left his Melbourne-based marketing job after 40 years and moved to Franklin on the Huon River to learn how to build traditional boats at The Wooden Boat Centre. In 20 months he and six other full-time boatbuilding students produced a 9.1-metre Huon pine motor sailer, *The Huon Kelly*, which Kelvin sponsored.

"It was the greatest learning experience of my life," he said, "Learning these new skills leaves marketing for dead."

He's staying at Franklin to pass on those skills to new students – and planning a circumnavigation of Tasmania in his new boat.

Tahune AirWalk, Forestry Tasmania's ecotourism experience on the banks of the Huon River below its junction with the Picton.

The 597-metre walk rises to 20 metres above the trees, while the cantilever at the end is nearly 50 metres above the level of the river.

Southport Lagoon

Recherche Bay

On the south coast, the geological upheavals that heralded the formation of Gondwana carved the face that Tasmania and its offshore islands present to the world today. The face is at its most dramatic on the south and south-east coasts and at its most welcoming in the long coastal plains of the Furneaux Group.

Beyond the capes of the south coast, there are small vestiges of the land mass that kept their heads above water after the last Ice Age, Mewstone Rock and Pedra Branca. They are individually significant as wildlife habitats – the Mewstone for albatrosses and Pedra Branca for its unique skink.

Pedra Branca, 26 kilometres off the Tasmanian mainland, looks like a static ocean liner or a tenement building sculpted from stone. Each deck of its terraces is home to different members of its community of white-capped albatrosses, gannets, black-faced cormorants, silver gulls, petrels, penguins, fur seals and the skinks.

There are six separate colonies of the skink on the island. They shelter in deep crevices and cracks which provide essential protection from wind, salt and spray, and feed on small invertebrates like insects and spiders and fish regurgitated from seabirds. They live as long as 15 years.

The Mewstone is a mitre-shaped rock that few are privileged to scale. On the island is a huge colony of white-capped albatrosses, fairy prions, petrels and cormorants. Each autumn new albatross chicks, as big as fully-fledged silver gulls, sit atop their tall thrones of nests, waiting to be fed by their parents. Today there is a conservation program to monitor the flight of the chicks once they have left the rock and headed out to sea. Birds from the Mewstone, with green blazes sprayed across their chests, have been reported off the coast of Africa.

Out to the west is Maatsuyker Island, the second largest in this group (De Witt is the largest). The Maatsuyker lighthouse was built in 1891 and continues to be manned for weather observations although the light is now automated. The keepers are resupplied by helicopter. Like most of the islands off the Tasmanian coast, there is a frequent problem with low cloud. Sometimes, it is necessary for the helicopters, after taking off from the top of the island near the keepers' cottages, to shimmy down the face of the island until the sea becomes visible.

The south coast is at its most striking in the hours before dawn and at dusk. At nightfall, as you absorb their grandeur on the passage back to the Tasmanian mainland from the islands, the capes lose their colour in the failing light but they retain their definition, zones of deepening grey.

Eddystone Rock rises 30 metres from the Southern Ocean east of Pedra Branca. The rock was named by Captain James Cook in 1777 because of its apparent resemblance to Eddystone lighthouse off the coast of Cornwall in the UK.

Eddystone Rock

Pedra Branca (which means 'white rock' in Portuguese) resembles an ocean liner off the Tasmanian south coast. It is an important seabird rookery and also has a unique skink. In April 2005 New Zealand scientist Hamish Saunders, on the island to study its fauna, was swept from the rock by a huge wave and was never found.

Surprise Bay with islands of the Maatsuyker Group in the distance

A column of sedimentary rock between the vertical forms of dolerite

The Mewstone

A shy albatross fledgling on Mewstone Rock

Hobart scientist Dr Graham Robertson has made the study and preservation of the albatross his life's work.

Ile du Golfe

Needle Rocks at Maatsuyker

Louisa Island. A sand spit provides access to Louisa Bay at low tide.

De Witt Island

Cox Bluff

Cox Bight looking east to the Ironbound Range

One of Tasmania's best known yachts, *Solandra*, off the south coast

Stephens Bay with the East Pyramids in the distance

Morrie Wolf has been fishing off the south and west coasts for more than 40 years. Today he and his wife Christine run a charter business out of Kettering with their boat *La Golondrina*.

A midden at Stephens Bay showing the diverse range of Aboriginal food

Breaksea Islands guarding the entrance to Port Davey

Port Davey

Bathurst Harbour and Mt Rugby

Kayaking in Bramble Cove

Balmoral Beach

Mouth of Spring River

Balmoral Beach

Strahan

Gordon River

Rafting the Franklin River, which flows into the Gordon

The West Coast Wilderness Railway turntable at Queenstown. The rack and pinion steam railway ran from 1896 until 1963 between the Mount Lyell copper mine at Queenstown and the port of Strahan. The line has been restored and is operating again as a tourist attraction. It runs for 35 kilometres across 40 bridges through spectacular scenery.

Queenstown

The Casey Joneses of the West Coast Wilderness Railway – Geoff Haines and Russell Francis

Huon pine millers on the West Coast maintain
a tradition that goes back to the early days of
European settlement.

Sculptor Greg Duncan with a Huon pine mural at the Wall in the Wilderness, his massive gallery at Derwent Bridge that will eventually extend 100 metres into the bush towards Lake St Clair. It will house his huge Huon pine timber sculptures, his works in bronze and stone – works that tell the epic history of forestry and hydro development in Tasmania, the industries that, together with mining, form the backbone of our development.

Paul Lees, Yvonne Gaffney, Lester Shadbolt, Alex Simpson and John Stevenson at "The Retreat" at Pieman Head

The cleanest air in the world blows steadily and strong on Cape Grim where an official air monitoring station keeps vigil on its quality.

Chris Richard, Stuart Baly and Jill Cainey checking the aerosol samples and flasks on the deck at Cape Grim

King Island

Potter and artist Caroline Kininmonth first visited King Island in 1989, saying she was drawn to the "intense fertility" that surrounded her on the land and in the sea. Her work is displayed at the brightly-coloured Boathouse Gallery in Currie harbour. She works from a pottery on Wharf Road.

Stonemason Jim Scott at the pottery, the former printery on the island

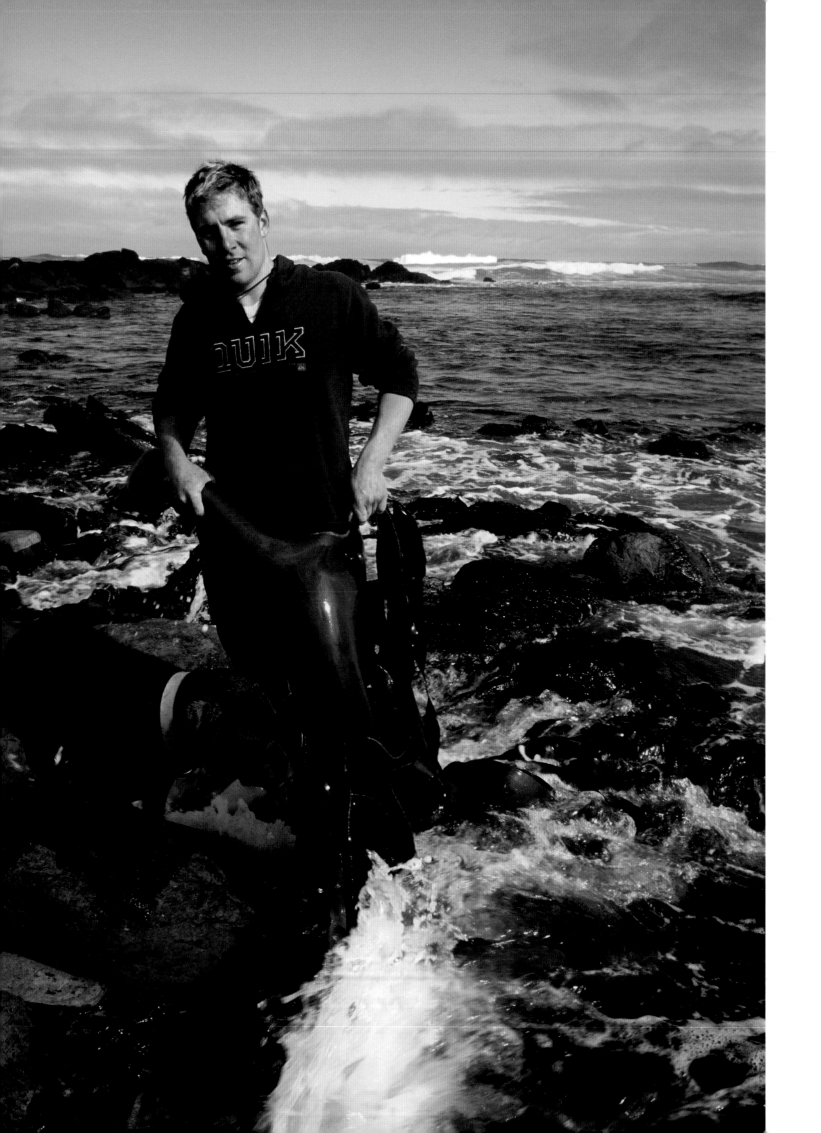

Pennys Lagoon, a rare, suspended freshwater lagoon near Cape Wickham

Kelper Evan Bott at Stokes Point. The storms that pound King Island provide a regular crop of bull kelp that is harvested by Kelp Industries, formed on the island in 1975. The kelp, which is dried and crushed once harvested, is said to be the richest source of alginates in the world. Alginates are used to thicken products such as ice cream, toothpaste, paper, paint and cosmetics. The bull kelp around King Island can grow up to 14 cm per day, making it the most prolific source in the world.

An ancient forest, calcified when it was covered by sand, is exposed in its unusual new form on the south of King Island.

Pearshape Lagoon

The wharf at Naracoopa, from where local mineral sands rutile and zircon were shipped for processing in the US

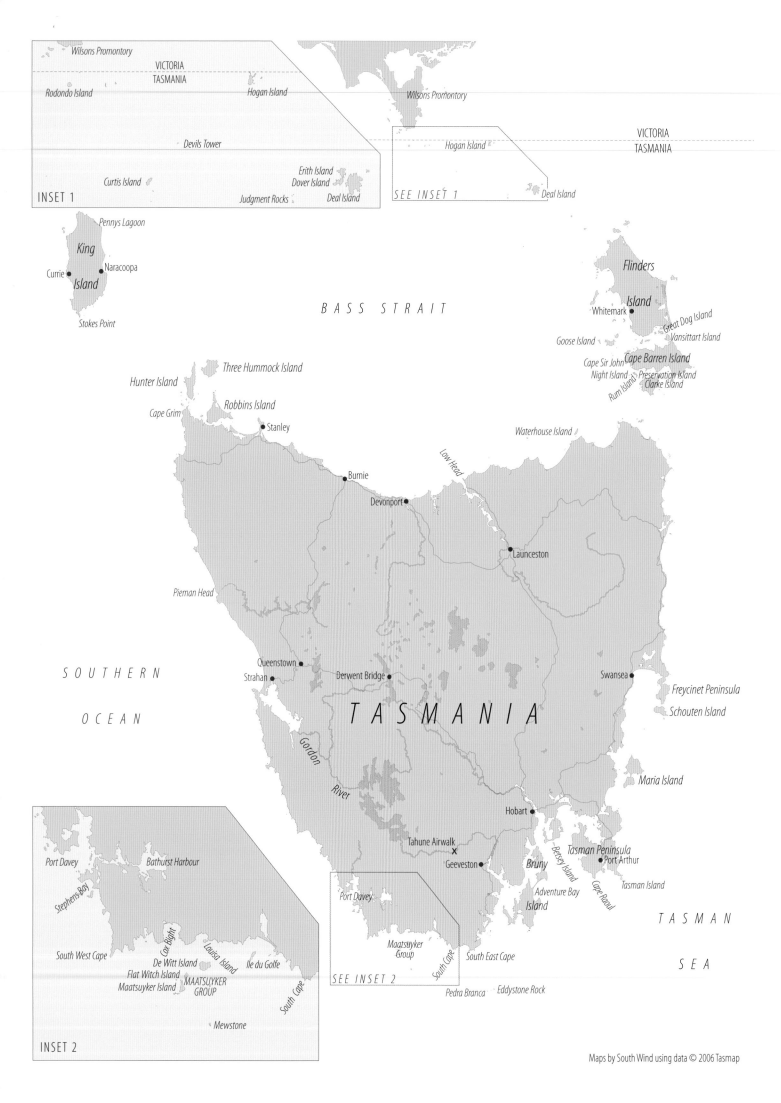

INSET 1

Wilsons Promontory

VICTORIA
TASMANIA

Rodondo Island

Hogan Island

Devils Tower

Erith Island
Dover Island

Curtis Island

Judgment Rocks

Deal Island

SEE INSET 1

Wilsons Promontory

Hogan Island

VICTORIA
TASMANIA

Deal Island

Pennys Lagoon

King

Currie • • Naracoopa

Island

Stokes Point

Flinders

Island

Whitemark •

Great Dog Island

Goose Island

Vansittart Island

Cape Sir John *Cape Barren Island*

Night Island *Preservation Island*

Rum Island Clarke Island

B A S S   S T R A I T

Three Hummock Island

Hunter Island

Robbins Island

Cape Grim • Stanley

Waterhouse Island

Low Head

• Burnie

Devonport •

S O U T H E R N

O C E A N

Pieman Head

• Launceston

Queenstown •

Strahan •

Derwent Bridge •

Swansea •

Freycinet Peninsula

Schouten Island

T A S M A N I A

Gordon

River

Maria Island

Hobart •

Tahune Airwalk
✕

Geeveston •

Bruny

Tasman Peninsula

Port Arthur •

Betsey Island

Island

Adventure Bay

Cape Raoul

Tasman Island

T A S M A N

INSET 2

Port Davey

Bathurst Harbour

Stephens Bay

South West Cape

Cox Bight

Louisa Island

De Witt Island

Ile du Golfe

Flat Witch Island

MAATSUYKER

Maatsuyker Island GROUP

Mewstone

South Cape

SEE INSET 2

Port Davey

Maatsuyker
Group

South Cape

South East Cape

Pedra Branca Eddystone Rock

S E A

Maps by South Wind using data © 2006 Tasmap

152

ACKNOWLEDGEMENTS

Premier Paul Lennon

The Hon. William J. E. COX AC, RFD, ED for the foreword

Bruce Montgomery for the text

Vicki Montgomery for editing

Pauline de Vos for design

Alice Bennett for photography and creative input

Peter Boyer and Tasmap for the map

Printing Authority of Tasmania – Ian Rosevear Chris Goodluck

Department of Premier and Cabinet – Julie Pellas

Gordon River Cruises – Alan Gifford Graeme Ridler

Tasair – Ralph Schwertner Ian Holmes Matthew Bester Henry Ellis

Par Avion – Greg Wells Marty Scott

Roaring Forties Ocean Kayaking – Ian Balmer Kim Brodlieb Toby Story Judd Hill Jerry Romanski

Resolution Adventures – Chris Fenner Judy Clark

Cape Barren Island Community – Sue Summers Sandra Reid Knud Andersen

Mures – Jill Mure Will and Judy Mure

Fujifilm – Kevin Cooper Graham Carter

Healing Dreams Retreat – Liz Frankham

Kelp Industries Pty Ltd – John Hiscock

Southern Ocean Adventures – Dave Wyatt

Raptor and Wildlife Refuge of Tasmania Inc. – Craig Webb

The Wall – Greg Duncan

Department of Primary Industries, Water and Environment – Dr Rosemary Gales Alex Simpson

Cape Grim Baseline Air Pollution Station – Jill Cainey Chris Richard Shane McEwan Stuart Baly

West Coast Wilderness Railway – Geoff Haines Russell Francis

Evan Bott Craig Broadbent Daryl Butler Bernice Condie Arne Eriksson Brodie Fazackerley
Yvonne Gaffney Darryl Gerrity Teresa Derrick Ian Hall Robert Hall Tony Harper Rex Johnson Jr
Caroline Kininmonth Paul Lees James and Lindsay Luddington Shai Maynard Alison McGregor
James McGregor Stuart and Adrienne McGregor Roger Murphy Mike Nichols Rob Pennicott
Ian and Gail Plowman Graham Robertson Jim Scott Lester Shadbolt
Gundars Simsons John Stevenson Fiona Stewart Morrie Wolf